YOUNG SCIENTISTS EXPLORE
ROCKS & MINERALS
Book 11 — Intermediate Level
By **Jerry DeBruin**

TABLE OF CONTENTS

Title	Page
Preface: A Note to Teachers and Parents	ii
Overview	1
Master Independent Study Contract	2
Young Scientists Explore the Western World of Rocks and Minerals	3
Young Scientists Explore the Eastern World of Rocks and Minerals	4
Geologists Study the Old	5
And Discover the New	6
The Planets	7
Planet Earth's Moon	8
Getting to the Core of Planet Earth and Its	9
Drifting Continents	10
Planet Earth Sometimes Quakes	11
And Rumbles	12
Geological Wonders of the World	13
Geological Wonders of the United States	14
Facing Rocks Near Your Community	15
And Downtown	16
Rocks and Minerals at Work	17
And at School	18
Tiny Rock Crystals Make	19
Large Crystal Gardens	20
Classify Rocks Into	21
A Pocketful of Rocks	22
Collect and Test Rocks	23
You Crack Me Up . . . In More Ways Than One	24
An Ending, a Beginning	25
Begin to Plan a Career in Geology	26
Teacher/Parent Guide	27

illustrated by Jeane Swemba cover by Vanessa Filkins and Tom Sjoerdsma Copyright © Good Apple, Inc., 1986
ISBN No. 0-86653-341-9 Printing No. 98765 Good Apple, Inc., Box 299, Carthage, IL 62321-0299
The purchase of this book entitles the buyer to reproduce student activity pages for classroom use only. Any other use requires written permission from Good Apple, Inc.
All rights reserved. Printed in the United States of America.

PREFACE
A NOTE TO TEACHERS AND PARENTS

Dear Friends,

Thank you for the many kind comments about my previous books, *Creative, Hands-On Science Experiences* (CHOSE) published by Good Apple, Inc., in 1980 and the first nine books in the *Young Scientists Explore* series published by Good Apple, Inc., in 1982, 1983 and 1985.

This book, *Young Scientists Explore Rocks & Minerals*, is written as a result of your comments and suggestions, comments that suggested a need for handy science books that contain activity sheets which could be duplicated for students' use. *Young Scientists Explore Rocks & Minerals* is an attempt to meet this need.

Young Scientists Explore Rocks & Minerals is the eleventh in a series of books that features science activity sheets which provide youngsters with a variety of interesting and challenging science-related experiences using experiments, puzzles, cutouts, hidden pictures, and charts. Youngsters are encouraged to gain firsthand experience while doing the activities, for it is through firsthand experience that meaningful learning occurs. Emphasis throughout the book is placed on the development of creativity in the learner. It is hoped that this creativity will be nurtured both during and after the activities are done.

You, as teachers and parents, are encouraged to make duplicating masters, copies and transparencies from the pages found in *Young Scientists Explore Rocks & Minerals*. The activities, written for various grade levels, are interdisciplinary and nonsequential in nature and are meant to integrate and extend existing science instruction with other academic areas. To achieve this purpose, the book features a contract system, activities and a Teacher/Parent Guide that supplies background information, including the title, major concept and materials needed for each activity plus a brief description on how to teach the activity with possible answers. When completed and mounted on single-layered cardboard, such as a file folder, and bound with tape, activity sheets 3 and 4 make up both the front and back covers of the Student Booklet. Completed student activity sheets can then be inserted between these covers and placed in a three-ring binder or file folder with the result being a permanent booklet for each youngster.

It is hoped that these activity sheets will touch many minds, hearts and hands and that much personal growth is experienced by all who do the activities. Keep in touch. Let me know how you are doing. It is always good to hear from you. Until then, best wishes in your continued growth as a scientist and as a complete human being.

Sincerely,

Jerry DeBruin

P.S. The author thanks Mr. David Link, Earth Science Specialist, Port Clinton City Schools, Port Clinton, Ohio, for his kind advice and helpful hints used in the writing of this book.

OVERVIEW: Use contracts to explore the world of rocks and minerals.

The contract system used in this book features "hands-on" science activities that are expanded upon by youngsters when involved in science. The advantages of using contracts are many. Some of these advantages include:

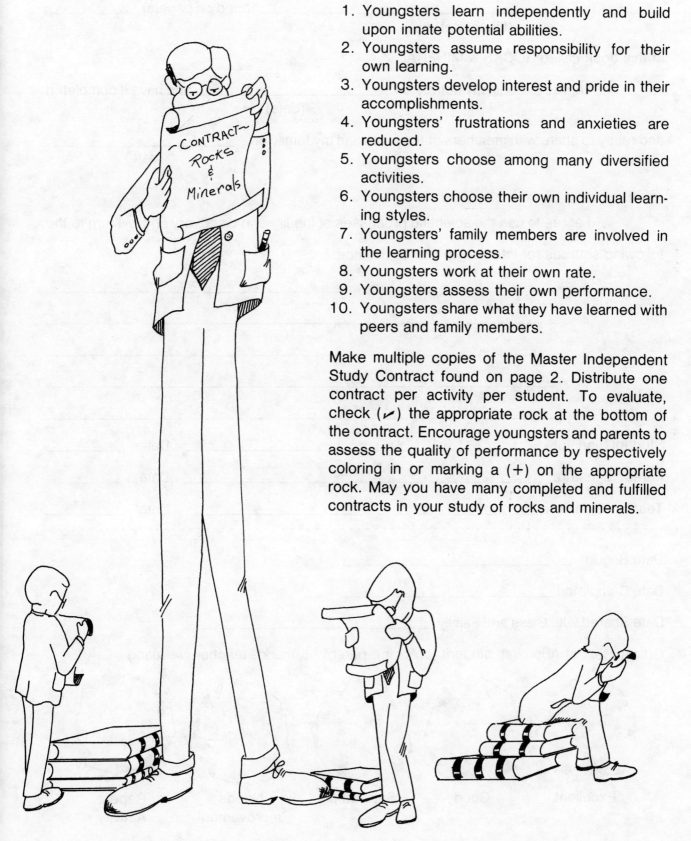

1. Youngsters learn independently and build upon innate potential abilities.
2. Youngsters assume responsibility for their own learning.
3. Youngsters develop interest and pride in their accomplishments.
4. Youngsters' frustrations and anxieties are reduced.
5. Youngsters choose among many diversified activities.
6. Youngsters choose their own individual learning styles.
7. Youngsters' family members are involved in the learning process.
8. Youngsters work at their own rate.
9. Youngsters assess their own performance.
10. Youngsters share what they have learned with peers and family members.

Make multiple copies of the Master Independent Study Contract found on page 2. Distribute one contract per activity per student. To evaluate, check (✓) the appropriate rock at the bottom of the contract. Encourage youngsters and parents to assess the quality of performance by respectively coloring in or marking a (+) on the appropriate rock. May you have many completed and fulfilled contracts in your study of rocks and minerals.

Master Independent Study Contract

I _____ agree to complete the activity entitled
(name)

_____ found on page(s) _____
(title of activity)

in this book called ROCKS & MINERALS.

I agree to begin my activity on _____ and have it completed
(date)

and ready to share with members of the class and my family on _____.
(date)

I agree to use the scientific processes of inquiry and discovery and will turn to the following sources for information and assistance.

1. _____
2. _____
3. _____
4. _____
5. _____

Student Signature _____ Date _____

Parent Signature _____ Date _____

Teacher Signature _____ Date _____

- -

Date Begun _____

Date Completed _____

Date Shared with Class and Family _____

Official Seal of Approval: student (color in), parent (+ mark), teacher (✓ mark)

| Excellent | Good | Satisfactory | Needs Improvement | Repeat Activity |

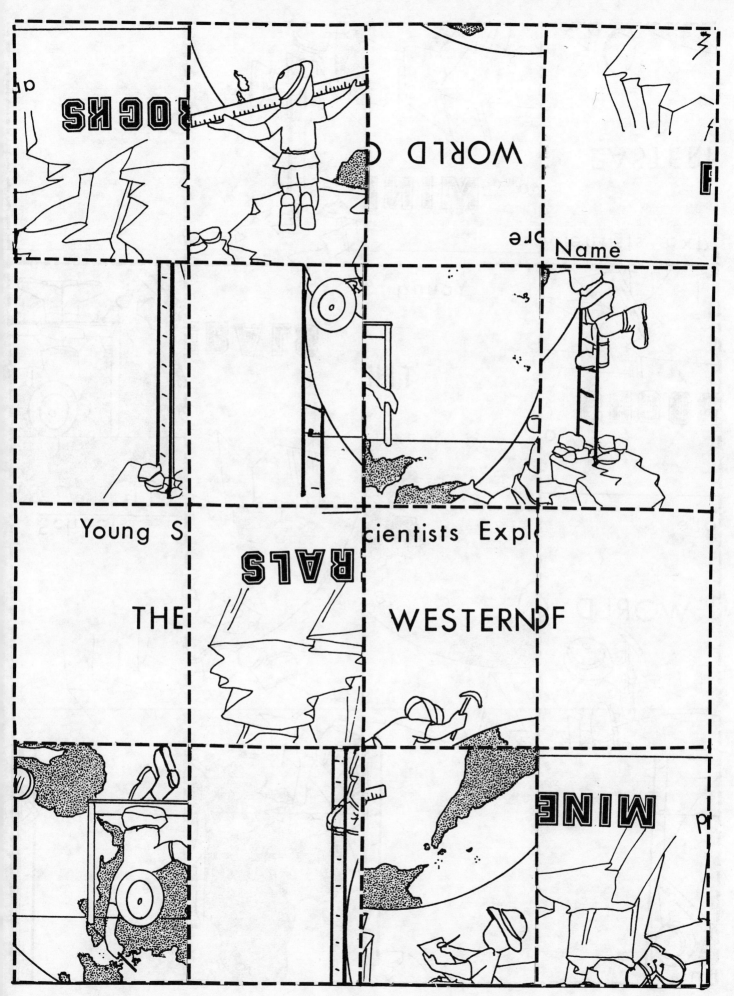

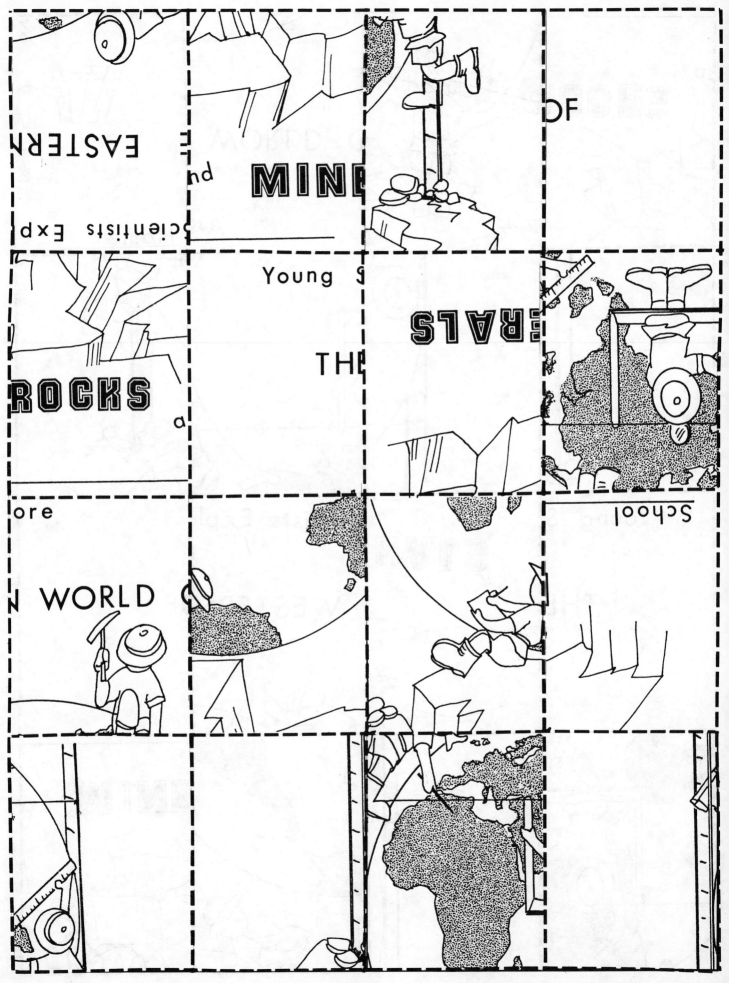

GEOLOGISTS STUDY THE OLD...

Geology is the study of the earth. Scientists called geologists study how the earth was formed and how it changes. You can be a geologist, too. Cut apart the cards below and glue to individual tagboard squares. Rearrange the squares, in order, to show the history of our universe. Glue the squares to a large sheet of cardboard, showing the correct sequence of events. Then find out the approximate time in years that it took for each event to occur. Study the history of each event; then write a newspaper article on what you think the next big event in history will be.

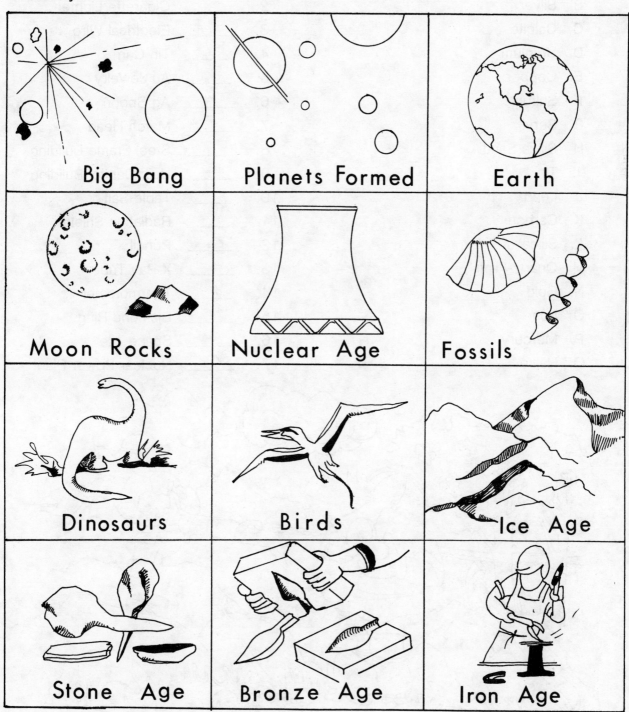

Big Bang | Planets Formed | Earth
Moon Rocks | Nuclear Age | Fossils
Dinosaurs | Birds | Ice Age
Stone Age | Bronze Age | Iron Age

AND DISCOVER THE NEW

Write the letter of the rock or mineral in Column A that best matches the object made from that rock or mineral found in Column B. The first one is done for you. Then, in the picture below, find and circle ten hidden objects named in Column B that are made from rocks and minerals. Find these objects in your home.

Column A

- A. Graphite
- B. Silver
- C. Calcite
- D. Uranium
- E. Copper
- F. Sulfur
- G. Beryl
- H. Silica Sand
- I. Tin
- J. Lead
- K. Carbon
- L. Silica
- M. Quartz-Flint
- N. Gold
- O. Iron
- P. Mercury
- Q. Halite

Column B

1. __H__ Drinking Glass
2. _____ Cigarette Lighter
3. _____ Electrical Wire
4. _____ Tin Can
5. _____ Au Jewelry
6. _____ Ag Spoon
7. _____ Match Head
8. _____ Steel Frame Building
9. _____ U.S. Capitol Building
10. _____ Table Salt
11. _____ Radiation Shield
12. _____ Pencil
13. _____ X-Ray Tube
14. _____ Thermometer
15. _____ Diamond Ring
16. _____ Space Tile
17. _____ Nuclear Power Plant

The Planets

Some scientists believe that many years ago our sun, earth and planets were a cloud of cold dust particles going through empty space. Gradually, the particles came together and formed a huge, spinning disk. As it spun, the disk became hot. The center of the disk became the sun. The planets were made from fiery gases and liquid matter near the sun. Our earth, a huge ball of rock, is one of these planets. Below is a picture of our solar system, located in the Milky Way galaxy. Name the planets. Glue page to a strong piece of cardboard. Find a rock or pebble that matches the size of each planet. Line up rocks according to size. Which planet is the largest? Smallest? Then tape rocks to matching planets on the cardboard. Find out what elements make up each planet. Report your results to the class.

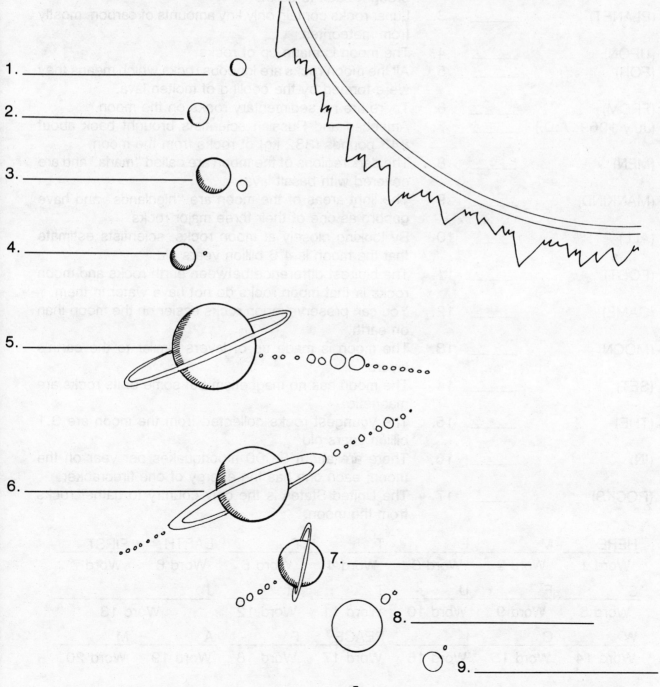

1. _____
2. _____
3. _____
4. _____
5. _____
6. _____
7. _____
8. _____
9. _____

Planet Earth's Moon

On July 20, 1969, Apollo 11 landed safely on planet Earth's satellite, the moon. From Apollo missions, scientists have identified various moon rocks and minerals. Read each statement below. In the blank, mark (+) if the statement is true, (−) if it is false. Then use the key words in ( ) in front of the **true** statements to decipher the secret message on the plaque left on the moon by the astronauts. Word numbers 1, 6, 7 and 17 are done for you below. Write the remaining words off the plaque to learn the message.

KEY WORDS

(THE) _____ 1. Rocks on the moon do not contain living organisms.

(WE) _____ 2. Scientists did not find fossils in the lunar rock samples brought back to earth.

(PLANET) _____ 3. Lunar rocks contain only tiny amounts of carbon, mostly from meteorites.

(UPON) _____ 4. The moon is made up of rocks.

(FOR) _____ 5. All the moon rocks are igneous rocks which means they were formed by the cooling of molten lava.

(FROM) _____ 6. There are no sedimentary rocks on the moon.

(July 1969, A.D.) _____ 7. American and Russian scientists brought back about 843 pounds (382 kg) of rocks from the moon.

(MEN) _____ 8. The dark regions of the moon are called "maria" and are covered with basalt lava.

(MANKIND) _____ 9. The light areas of the moon are "highlands" and have gabbro as one of their three major rocks.

(ALL) _____ 10. By looking closely at moon rocks, scientists estimate that the moon is 4.6 billion years old.

(FOOT) _____ 11. The biggest difference between earth rocks and moon rocks is that moon rocks do not have water in them.

(CAME) _____ 12. You can preserve moon rocks easier on the moon than on earth.

(MOON) _____ 13. The moon is made up of layers similar to the earth's layers.

(SET) _____ 14. The moon has no magnetism but some of its rocks are magnetic.

(THE) _____ 15. The youngest rocks collected from the moon are 3.1 billion years old.

(IN) _____ 16. There are about 3,000 moonquakes per year on the moon; each one has the energy of one firecracker.

(ROCKS) _____ 17. The United States is the only country to gather rocks from the moon.

HERE	M	F	T	P	EARTH	FIRST
Word 1	Word 2	Word 3	Word 4	Word 5	Word 6	Word 7
S	F	U	T	M	J	
Word 8	Word 9	Word 10	Word 11	Word 12	Word 13	
W	C	I	PEACE	F	A	M
Word 14	Word 15	Word 16	Word 17	Word 18	Word 19	Word 20

8

GETTING TO THE CORE OF PLANET EARTH AND ITS...

Planet Earth, the only known planet where life exists, is the sun's third planet. The center of the earth lies about 4,000 miles (8,400 km) beneath our feet. Thus far, however, we have drilled only about 7 miles (11 km) into the earth itself. Below is a picture of the earth and its parts. Unscramble the letters. Write the correct name of each part in the blank. Color each part. Then write the words *graphite* and *basalt* on the crust, *iron* and *magnesium* on the mantle and *nickel-iron* in the core of the earth. These rocks and minerals make up various parts of the earth. In your science log, write a story about how drilling into the earth could someday heat your home or predict an earthquake.

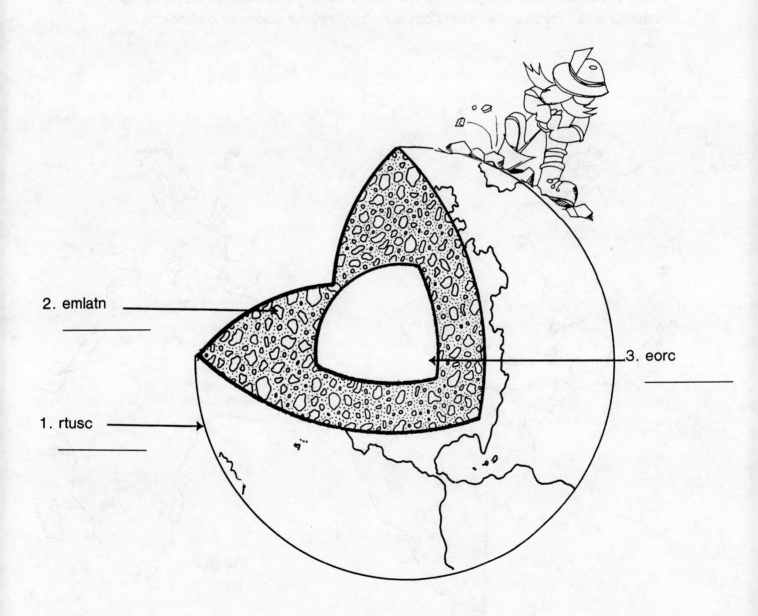

2. emlatn

3. eorc

1. rtusc

Drifting Continents

In the past 80 years, there has been much interest in the shapes of large land areas on earth and how they appear to fit together. Cut out the land areas, North and South America (Figure 1) and Africa (Figure 2), below to learn more about this idea. Glue shapes to thin pieces of cardboard. Cut out and label the continents. Slide Africa over to South America until you get a close fit. Then slide the northwestern coast of Africa along the eastern coast of North America to obtain a close fit. On a sheet of paper, trace around the "fitted together" Figures 1 and 2. You will have made a picture of how the continents may have fit together at one time in history and then drifted apart. Look up "continental drift." Record the results of your study in your science notebook.

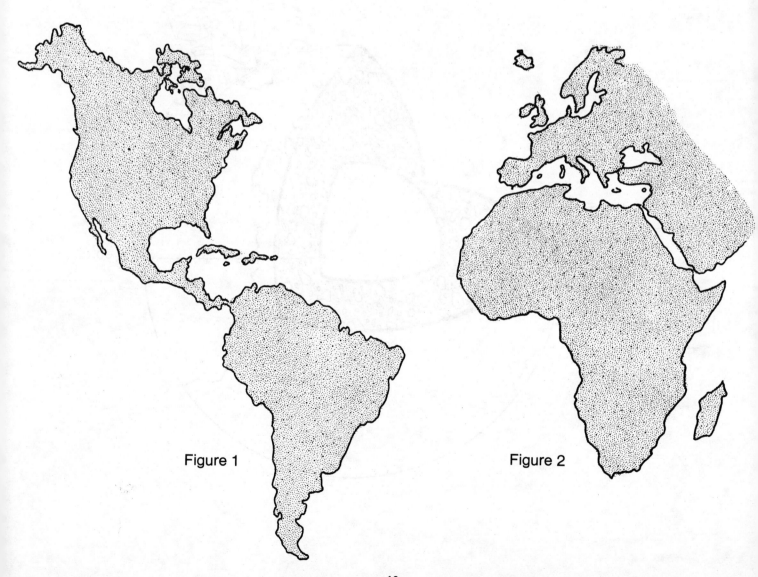

Figure 1

Figure 2

Planet Earth Sometimes Quakes...

During an earthquake, forces inside the earth cause the earth's crust to bend, break and snap into a new position. This is called a fault at which time rocks split, turn over and move around. You can do the same. Obtain a flat, nearly square rock. Stick a piece of tape to one corner of the rock. Note location of tape. Turn rock one **quarter** turn to the left. Note location of tape. Turn rock one **half** turn to the right. Observe location of tape. Is the location the same as when you began? Observe the picture of granite in row one below. Look over the other pictures of granite until you have found one that has been turned one **quarter** turn to the left, then one **half** turn to the right. Circle it. Do the same for rocks B, C, and D. How are they the same? Circle it. Do the same for rocks B, C, and D. How are they the same? Different? When finished, lie on your back. Pretend that you are the rock granite in an earthquake. Roll one **quarter** turn to the left, then one **half** turn to the right. Upon what side of your body are you now resting? _____ Now, try being a rock that lies face down. Are your results the same or different?

  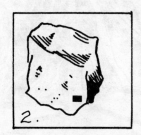 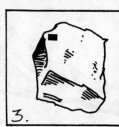 

A. GRANITE | 1. | 2. | 3. | 4.

B. OBSIDIAN | 1. | 2. | 3. | 4.

   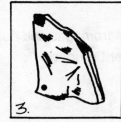 

C. SHALE | 1. | 2. | 3. | 4.

D. SLATE | 1. | 2. | 3. | 4.

And Rumbles

The earth's crust is made up of a layer of rocks that float on top of a heavy, hot, thick liquid called magma. Sometimes the hot magma bursts through the crust to form lava. The entire happening is called a volcano, shown in the picture below. Using the words below, label its parts. Then build your own volcano by following the directions below.

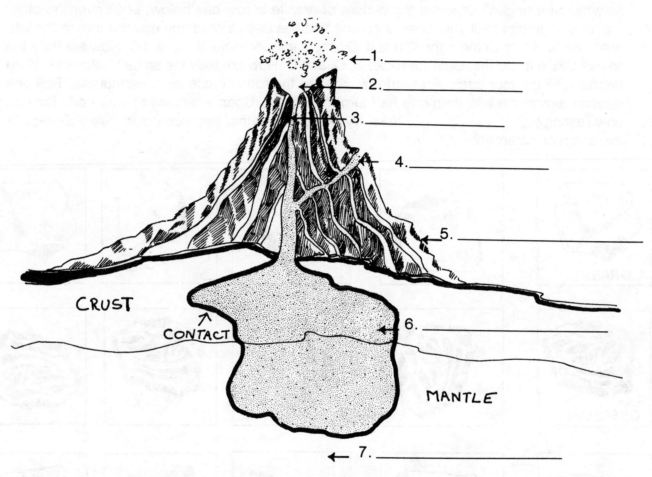

1. _____
2. _____
3. _____
4. _____
5. _____
6. _____
7. _____

Words to chose from: gas and dust (ejecta), lava flow, crater, conduit, liquid magma, hot rock magma, side vent eruption

Building a Volcano

You and your parents or teacher can make a volcano. You need a small juice can or a 35mm film can for the earth, a bucket or container for catching lava, bits of styrofoam for rocks, liquid soap, baking soda, vinegar, and red food coloring. Place empty juice or film can in lava catcher. Add styrofoam bits to bottom of can. Fill can ½ full with liquid soap. Add red food coloring for magma color. Mix in baking soda. Stir. Gradually add vinegar to mixture. Carefully observe the action of the magma as it carries the styrofoam bits (rocks) out of the volcano to form lava. Write a story in your science notebook about how your volcano is similar to that of Mt. St. Helens. Be sure to include facts from the picture above.

GEOLOGICAL WONDERS OF THE WORLD

Below is a map of the world. Letters in the boxes (□) show the locations of seven continents and three major oceans of the world. List these in the chart below. Numbers in circles (O) show the locations of famous geological features. Using the words below, make a chart on the other side of this page that looks like this:

Number	Geological Feature	Country	Purpose
1	Rosetta Stone	Egypt	Hieroglyphics script found here.

Then plan an imaginary trip to one of these sites. In your science log, describe your travel plans, what you might see and the results of your trip.

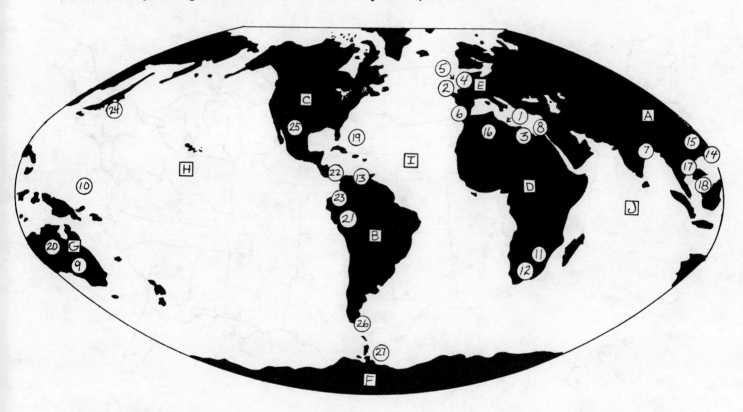

Words to choose from:
Rosetta Stone, Stonehenge, Great Pyramid of Cheops, White Cliffs of Dover, Blarney Stone, Rock of Gibraltar, Mt. Everest, Dead Sea, Lake Eyre, Mariana Trench, Victoria Falls, Tugela Falls, Angel Falls, Great Wall of China, Gobi Desert, Sahara Desert, Plain of Jars, Angkor Wat, Atlantis (possible), Ayers Rock, Cuzco, Tikal, Sangay Volcano, Mt. Fuji, Grand Canyon, Cape Horn, Weddell Sea

The seven continents of the world are:
A. _____
B. _____
C. _____
D. _____
E. _____
F. _____
G. _____

The three major oceans of the world are:
H. _____
I. _____
J. _____

13

GEOLOGICAL WONDERS OF THE UNITED STATES

Below is a map of the United States. Dots (•) on the map show the locations of famous geological features. Using the numbers below, write the number of the geological feature next to the dot that shows the location of that geological feature. Then on the other side of this page, write the name of the state in which that geological feature is found. Identify your favorite geological feature and tell why it's special to you.

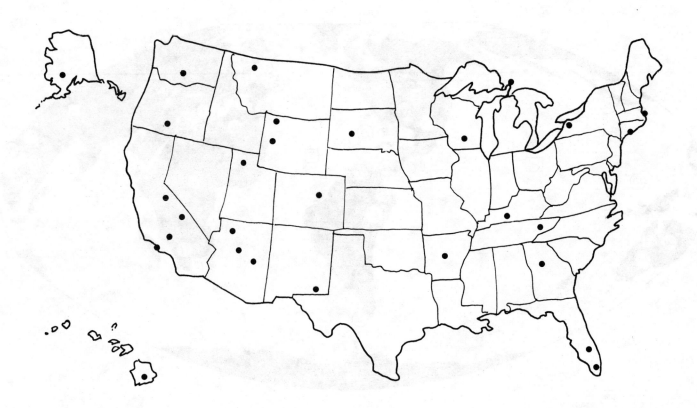

Words to choose from:
1. San Andreas Fault, 2. Mt. St. Helens, 3. La Brea Tar Pools, 4. Badlands, 5. Petrified Forest, 6. Grand Canyon, 7. Old Faithful, 8. Yosemite Falls, 9. Rocky Mountains, 10. Carlsbad Caverns, 11. Niagara Falls, 12. Great Salt Lake, 13. Lake Okeechobee, 14. Cape Cod, 15. Meteor Crater, 16. Cleopatra's Needle, 17. Great Smoky Mountains, 18. Everglades, 19. Mackinac Island, 20. Hot Springs, 21. Yellowstone, 22. Jackson Hole, 23. Mt. McKinley, 24. Hawaii Volcano House, 25. Crater Lake, 26. Death Valley, 27. Glacier National Park, 28. Mammoth Cave, 29. Wisconsin Dells, 30. Stone Mountain

Facing Rocks Near Your Community...

Look around your community where the faces of rocks exist. This may be near a road cutting, seashore or hill. Observe the face carefully. On the clipboard below, write the direction (NSEW) you are facing. Then make a sketch of an interesting part of the rock face. After you complete your sketch, carefully look again and answer the questions below.

Were the rocks in layers, blocks or both? _____
Were any rocks tilted in layers? _____
What colors could you see in the rocks? _____
Was the rock face wet, moist or dry? _____
Did the rock face have a smell? If so, what smell? _____
Were any plants growing in or near the rocks? _____

Take a colored photograph of the rock face. Compare it to your sketch. How are they different? Same? Then examine these faces made from rock. Name each. Name the memorial on which these faces are found.

FACES
1. _____
2. _____
3. _____
4. _____ ←Mt. _____

And Downtown

Towns and cities are good places to study stones. The picture below shows a number of buildings that may be found downtown in your community. There is something missing, however, as the pictures are left blank. With permission, visit your downtown. Observe the faces of buildings, streets and monuments. Look for marble, granite, limestone and man-made stones such as concrete, tiles and bricks. Draw in and name the stones on the buildings in each picture. Then complete the chart below.

My Name		Downtown Name			Date
Name of Building	Age of Building	Name of Stone Spotted	Color	Natural or Man-Made	How Used? Is Stone Used for Structure or Decoration?

16

ROCKS AND MINERALS AT WORK...

Some of our greatest riches are called metals. These are found deep in the earth's crust. Metals make jewelry, cars, trains, ships, aircraft, buildings, bridges and other things. One such metal is hematite, used in the making of iron. Connect the dots in the picture below to show the path hematite takes as it's made into iron. Then go on a metal hunt. Fill in the chart below with the results of your findings.

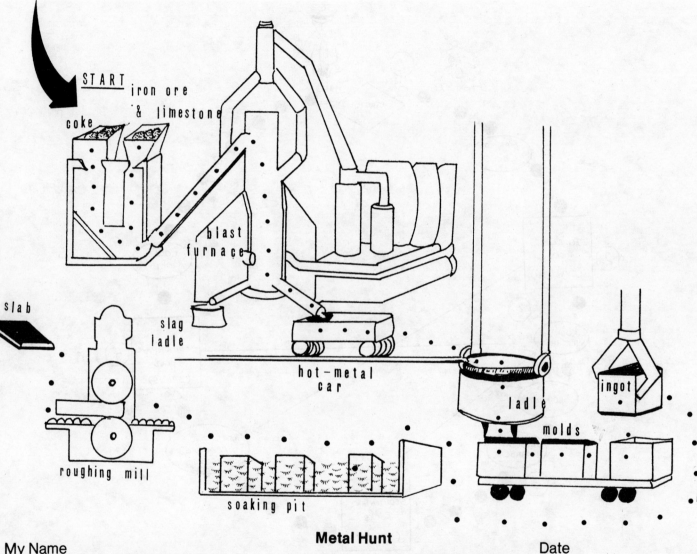

Metal Hunt

My Name _____   Date _____

Name of Metal	Object Found In	Color	Hard/Soft	Magnetic	Bends Using Fingers	Heavy/Light	Floats/Sinks

AND AT SCHOOL

The rock cycle goes on and on but at a very slow speed. When one type of rock is destroyed, another type of rock is formed. Study the picture of the rock cycle below. With a pencil, start at MAGMA. Connect the dots and arrows until you return to MAGMA. You will have traveled one complete rock cycle. Fill in the blanks below the picture with the word or words that tell what happens at each stop.

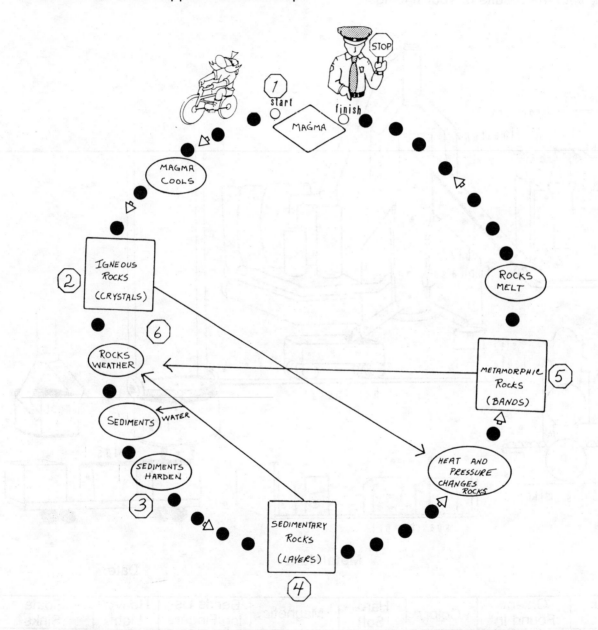

Stop 1: Start at MAGMA. Connect dots and arrows through MAGMA. Cools to make I _____ rocks at Stop 2. Continue on. Igneous rocks are weathered by agents of erosion such as wind, water, ice, chemical action, sun and gravity to form S_____ at Stop 3. Water mixes with sediments which then harden into S_____ rocks at Stop 4. Heat and pressure change sedimentary rocks into M_____ rocks at Stop 5. Metamorphic rocks melt and form M_____ again at Stop 1. BONUS: Metamorphic and sedimentary rocks can also be changed into S_____ by W_____ at Stop 6.

Tiny Rock Crystals Make...

Rocks are made up of many kinds of crystals. Some are colored, some colorless, some hard, some soft. In all, there are thirty-two different classes of crystals. In Circle A below, sprinkle a tiny amount of salt. With a magnifying glass, look carefully at the shape of the salt crystals. In Square B, draw the shape of the salt crystals. Figure C shows a spread out version of the thin layers of each side in a salt crystal. Cut out, fold and tape together the sides found in Figure C. Name the shape of a salt crystal. Then on a separate piece of paper draw on, cut out and put together the flat shape of a diamond crystal.

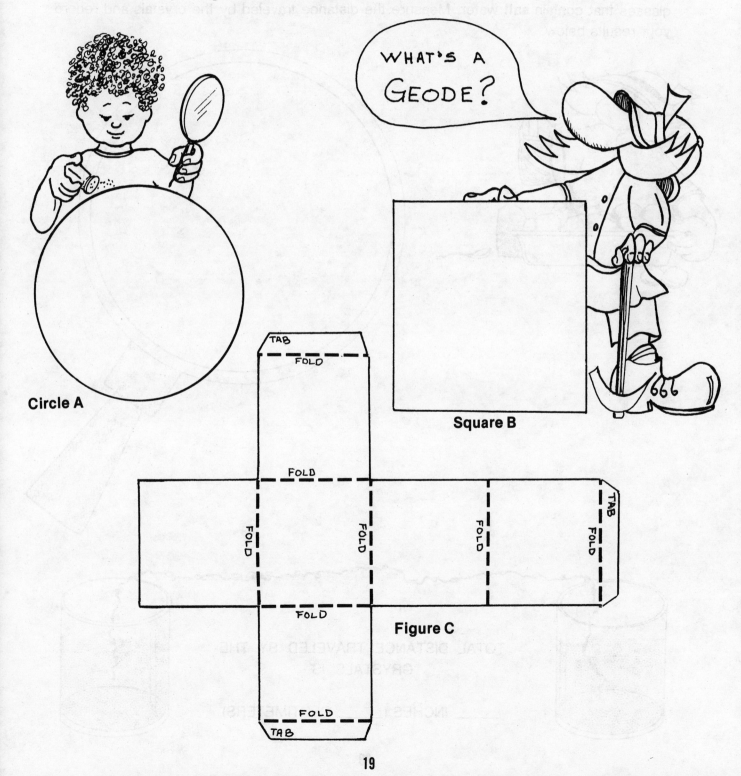

Large Crystal Gardens

To make a crystal garden, place several lumps of charcoal or coal in a glass or metal container. Mix 6 tablespoons (90 ml) of ammonia, 6 tablespoons (90 ml) of water, 6 tablespoons (90 ml) of salt and 6 tablespoons (90 ml) of laundry bluing in another container. Pour mixture over coal or charcoal. Add food coloring for colored crystals. Grow crystals on sticks and small branches. Study the designs and shapes made by the crystals. What geometric patterns do you see? In the magnifying glass below, make a drawing of the favorite part of your crystal garden. Color it. Then see if you can make crystals grow along a string between two glasses that contain **salt** water. Measure the distance traveled by the crystals and record your results below.

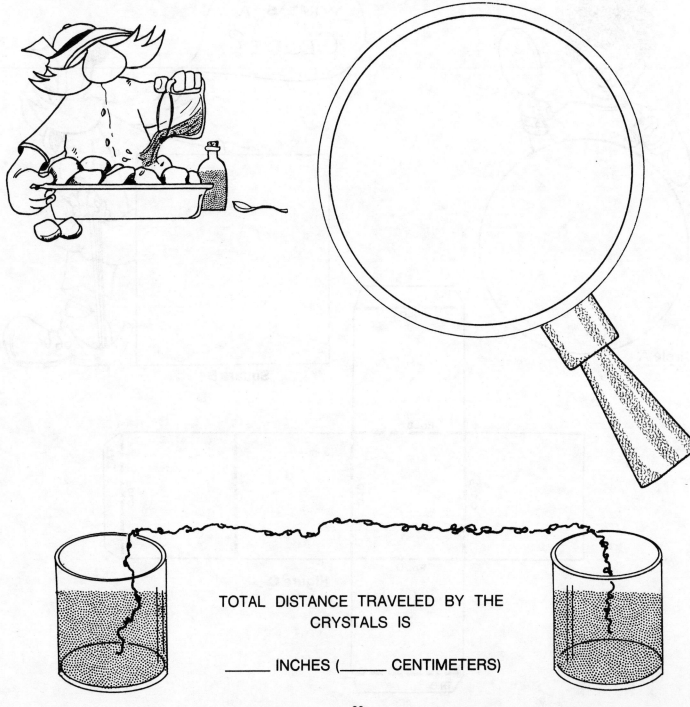

TOTAL DISTANCE TRAVELED BY THE CRYSTALS IS

____ INCHES (____ CENTIMETERS)

CLASSIFY ROCKS INTO...

Collect and study the twelve rocks below. Then cut out each picture on this page. Place the picture in the correct pocket from page 22. Rearrange the cards in each pocket so that the letters on each card, when placed in the correct order, will make a secret word that tells something about each class of rocks: Igneous, Metamorphic and Sedimentary. Record these secret words in the blanks below. Then gather other rocks and classify them in your collection.

IGNEOUS ROCKS (4)
_____
_____
_____
_____

METAMORPHIC ROCKS (4)
_____
_____
_____
_____

SEDIMENTARY ROCKS (4)
_____
_____
_____
_____

LETTERS (4)
__ __
__ __

LETTERS (4)
__ __
__ __

LETTERS (4)
__ __
__ __

SECRET WORD _____

SECRET WORD _____

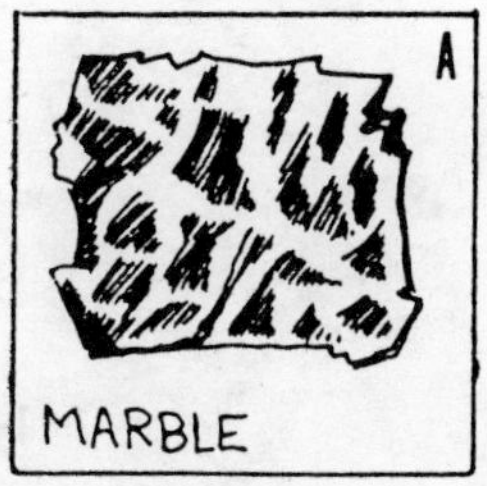

SECRET WORD _____

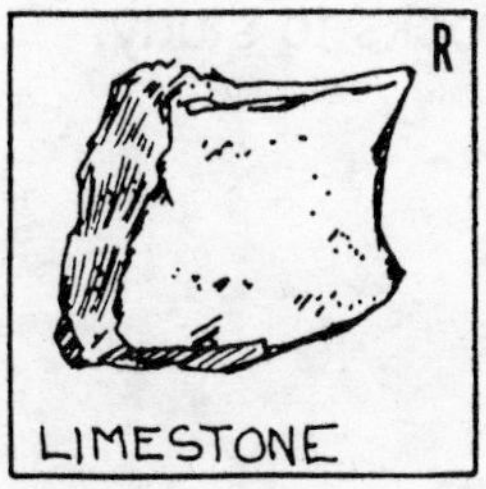

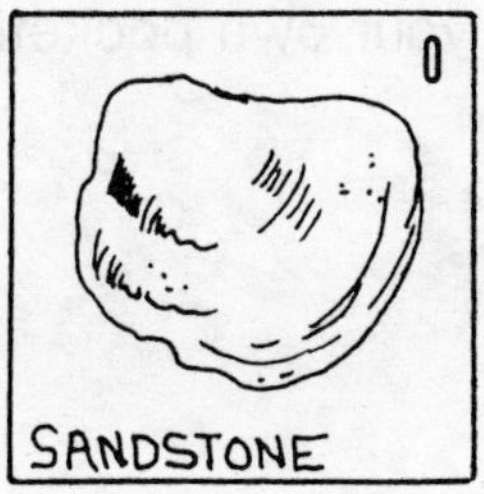

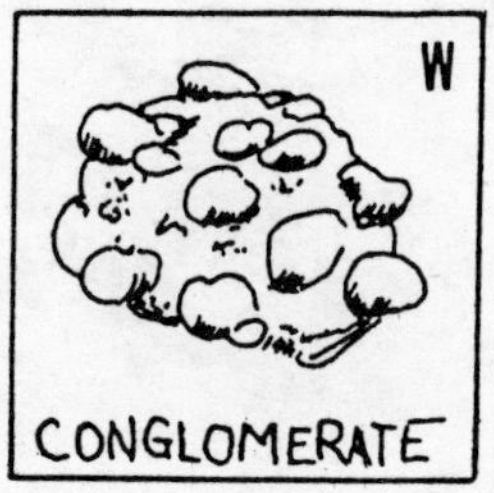

A — MARBLE
R — OBSIDIAN
N — SHALE
R — LIMESTONE
O — SANDSTONE
I — BASALT
H — SLATE
F — GRANITE
W — CONGLOMERATE
E — SCHIST
T — QUARTZITE
E — DIORITE

A POCKETFUL OF ROCKS

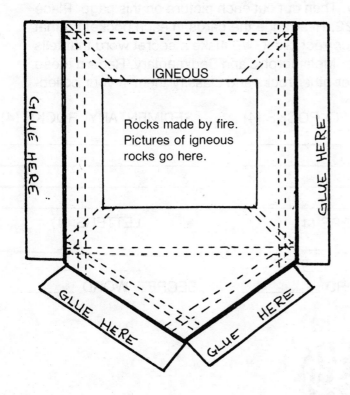

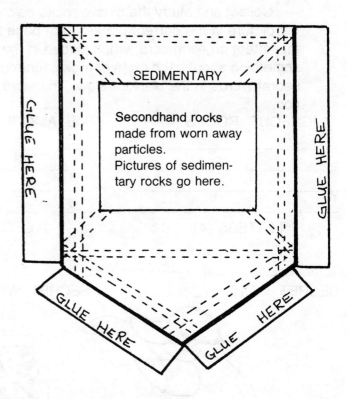

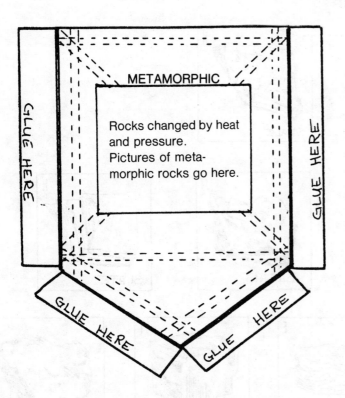

Here are drawings of three pockets that you can make to store the pictures of rocks found on page 21. The pockets are named after the three main classes of rocks: Igneous, Metamorphic and Sedimentary. Cut out each pocket. Fold on dark, solid lines. Glue to cardboard to make pocket card holder. Place matching pictures of rocks from page 21 in correct pockets. Then go outside and pick your own pocketful of rocks to study.

COLLECT AND TEST ROCKS

Go outside and find at least five small but different rocks. Using a piece of tape, label the rocks A, B, C, D, and E. Make a record of where and when each rock was found. Carefully clean each rock. Do tests on each rock. Fill in your results in the charts below. On a piece of cardboard, tape your rocks in order by using Mohs' Scale of Hardness as your guide. Then cut out the pictures of rocks below and put them in order according to hardness.

MY ROCK CHART

ROCK	COLOR		SHADE	WEIGHT	LAYERS	TEXTURE rough smooth	WRITES yes no	SHINY or DULL	CRYSTALS	VINEGAR ADDED	MOHS* RATING	SPECIFIC GRAVITY	OTHER
	dry	wet											
A													
B													
C													
D													
E													

MOHS' SCALE CHART

ROCK	Finger Nail	Penny	Nail	Knife	Glass	Tile	Color	Other
A								
B								
C								
D								
E								

*MOHS' SCALE OF RELATIVE HARDNESS

MATERIAL	WHAT IT WILL DO	Rating
TALC	so soft it's used for talcum powder	1
GYPSUM	a fingernail will scratch it	2
CALCITE	a copper penny will scratch it	3
FLUORITE	a steel knife will scratch it	4
APATITE	a knife scratches if you press hard	5
FELDSPAR	will scratch a knife blade	6
QUARTZ	will scratch glass (and all previous)	7
TOPAZ	will scratch quartz (and all previous)	8
CORUNDUM	will scratch all except a diamond	9
DIAMOND	will scratch everything	10

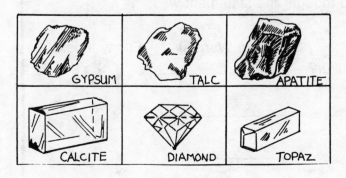

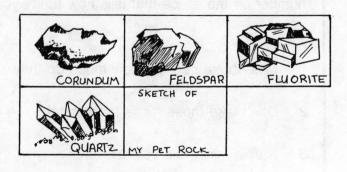

SKETCH OF MY PET ROCK

You Crack Me Up...

Write the letter of the picture in the blank next to the name that best tells the force that changes rocks into soil. Then, using a ruler, draw a one-inch (2.5 cm) vertical line in the box below starting at dot X. This line shows 1" (2.5 cm) of soil which took nature over 800 years to make.

It's a slow process but a valuable one as we need soil to grow crops to live. In your science log, make a sketch of how deep the topsoil and subsoil are where you live.

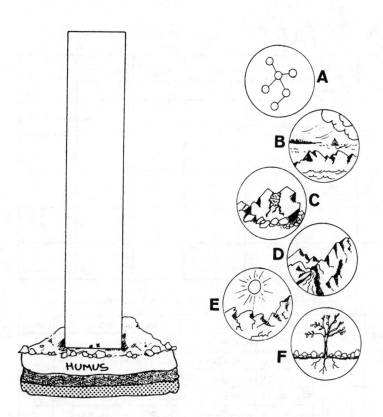

1. ____ Alternate Freezing and Thawing

2. ____ Freezing Water

3. ____ Chemical Reaction

4. ____ Wind and Water Wear Away Rocks

5. ____ Growing Plants

6. ____ Glaciers

in More Ways Than One

Sound-alike words are fun because they make puns that keep you thinking. In each sentence find and circle the sound-alike word(s) that sounds like the real word. Write the real word in the blank. Then on the 1-5 funny index (1 being most funny, 5 being funny), put a number on the space that tells how funny you thought the statement was.

	Alike Word(s)	Funny Index
1. I like those pencils. How much do they sulfur?	_____	
2. Don't take those rocks for granite!	_____	
3. A cold miner is a vein person.	_____	

AN ENDING, A BEGINNING

With permission from your teacher, parents and caretaker of a cemetery, visit a cemetery and study a tombstone that looks interesting to you. Tape a sheet of newsprint over the epitaph. Rub a crayon across the imprint so image can be seen on the newsprint. Then cut the newsprint to show the shape of the tombstone. Display. Complete your tombstone rubbing with information given below. Select one of the 16 symbols. Draw the symbol on the rubbing of your tombstone. Tell the class what you think your symbol means. Then write your own epitaph.

Symbols to choose from:
(1) Angel, (2) Finger Pointing Heavenward, (3) Opening Gates, (4) The Lamb, (5) The Holy Bible, (6) The Dove, (7) Flowers, (8) Hourglass, (9) Weeping Willow Tree, (10) Clasped Hands, (11) Tree Trunks, (12) Broken Flower, (13) Wreath, (14) Flower Bud, (15) Grapevine, and (16) Winged Cherubs

BEGIN TO PLAN A CAREER IN GEOLOGY

Are you interested in rocks and minerals? Earthquakes and volcanoes? Dinosaurs and fossils? If so, you may become a geologist, a person who studies the crust and rocks of the earth. Below are ten geological fields of study. Choose your career in geology by writing the letter of the field on the cap that best describes that field. Then write an employment ad for a newspaper listing your qualifications as a geologist.

A. Seismology  1. Study of layers of rock in the earth's crust.

B. Petrology  2. Study of fossils.

C. Stratigraphy  3. Study of earthquakes.

D. Oceanography  4. Study of igneous, metamorphic and sedimentary rocks.

E. Paleontology  5. Study of oceans and life in the oceans.

F. Hydrology  6. Study of minerals.

G. Sedimentology  7. Study of the movement and distribution of waters on the earth.

H. Planetology  8. Study of sediment and how they are deposited.

I. Mineralogy  9. Study of chemical and physical properties of the planets.

J. Meteorology 10. Study of weather.

26

TEACHER/PARENT GUIDE

YOUNG SCIENTISTS EXPLORE THE WESTERN WORLD OF ROCKS AND MINERALS
(front cover)

Activity Page Number 3

Concept: Identify geological features in the Western World

Tip(s): Have youngster cut activity sheet into designated 16 squares. Rearrange puzzle pieces to make complete picture. Glue pieces in place onto cardboard (back of writing tablets) or tagboard. Add name of student. Color. Decorate. Use as front cover for Student's Book. Discuss major features of the Western World including continents, countries and oceans. Duplicate activity page 4 for back cover of Student's Book. Follow same procedure. Add name of school. Discuss features of Eastern World. Laminate both covers. Tape front cover to back cover and use as a folder for pages that make a complete Student Book. Youngster can also use a file folder or a 3-ring binder for this purpose.

Answers: Completed project looks like this:

YOUNG SCIENTISTS EXPLORE THE EASTERN WORLD OF ROCKS AND MINERALS (back cover)

Activity Page Number 4

Concept: Identify geological features in the Eastern World

Tip(s): See Tips from Activity Number 3 above.

Answers: Completed project looks like this:

GEOLOGISTS STUDY THE OLD...

Activity Page Number 5

Concept: Identify stages in the history of the universe

Tip(s): Laminate activity cards for lasting durability. Extend activity to include the number of approximate years it took for each event to occur. Refer to *Dragons of Eden* by Carl Sagan for approximate time line.

Answer(s): Big Bang, Planets Formed, Earth, Moon Rocks, Fossils, Dinosaurs, Birds, Ice Age, Stone Age, Bronze Age, Iron Age, Nuclear Age

AND DISCOVER THE NEW

Activity Page Number 6

Concept: Many everyday objects are made from rocks and minerals.

Tip(s): After youngsters have completed the activity, discuss the possible uses for each object found in Column B; decrease intake of table salt to maintain health, use of drinking glass to make music are a few examples.

Answer(s): Part I—(1) H, (2) M, (3) E, (4) I, (5) N, (6) B, (7) F, (8) O, (9) C, (10) Q, (11) J, (12) A, (13) G, (14) P, (15) K, (16) L, (17) D. Part II—Ten hidden objects are circled below. These include: drinking glass, electrical wire, tin can, Au jewelry, Ag spoon, match head, table salt, pencil, thermometer, and diamond ring.

THE PLANETS

Activity Page Number 7

Concept: To identify the planets and how rocks played a role in their formation

Tip(s): Use two-sided, double-stick adhesive tape to glue rocks to cardboard. Order of planets: (1) Mercury, (2) Venus, (3) Earth, (4) Mars, (5) Jupiter, (6) Saturn, (7) Uranus, (8) Neptune, (9) Pluto. Remind youngsters, however, that Pluto now lies within the orbit of Neptune, thus Neptune is the farthest away planet. This will be true until the year 1999 when Pluto will resume its normal position. Youngsters can remember the usual order of the planets with this device: "Mary's Velvet Eyes Made John Stay Up Nights Pondering"; the current order: "My Very Elegant Motorcycle Jumped Sideways Upsetting Practically Nobody."

Answer(s): Planets according to size largest to smallest with approximate circumference in miles (km): Sun (2,717,446 mi. or 4,347,915 km), Jupiter (279,552 mi. or 447,283 km), Saturn (234,263 mi. of 374,820 km), Uranus (101,123 mi. or 161,797 km), Neptune (96,633 mi. or 154,621 km), Earth (25,133 mi. or 40,212 km), Venus (23,622 mi. or 37,795 km), Mars (13,263 mi. or 21,220 km), Mercury (9,527 mi. or 15,243 km), Pluto (5,857 mi. or 9,371 km).

PLANET EARTH'S MOON

Activity Page Number 8

Concept: Characteristics of moon rocks, Apollo 11 mission

Tip(s): To extend this activity, write to the National Aeronautics and Space Administration, (NASA) Washington, D.C. 20546 to obtain information on how you can become certified to obtain moon rocks, on loan, for your classroom use. Ask for a free copy of the booklet *What's New on the Moon* which contains excellent information on moon rocks.

Answer(s): All are true except #17 which is false. The Soviet Union has obtained lunar samples with its Luna 16 and 20 unmanned lunar missions.

Message on plaque: Here Men From The Planet Earth First Set Foot Upon The Moon, July 1969, A.D. We Came In Peace For All Mankind.

GETTING TO THE CORE OF PLANET EARTH AND ITS...

Activity Page Number 9

Concept: Parts of the earth

Tip(s): Assist youngsters in understanding the parts of the earth by using a cutaway apple, the peel being the crust, the pulp for the mantle and the core for the core of the earth. Discuss the rocks and minerals found in each part. Inform youngsters that the world's deepest hole is now 7 miles (11 km) beneath the Kola Peninsula, near the Arctic Ocean, and being drilled by the USSR.

Answer(s): (1) crust, (2) mantle, (3) core

DRIFTING CONTINENTS

Activity Page Number 10

Concept: Continental drift

Tip(s): Introduce activity by having youngsters let their bodies "drift" through space while listening to music. Complete exercise. Extend activity by making a transparency of the map below to show youngsters how all continents may have "fitted" together at one time in history. Inform youngsters that Africa and North America were once attached at the "overthrust belt" in the Appalachians forming one continuous plate, then drifted apart.

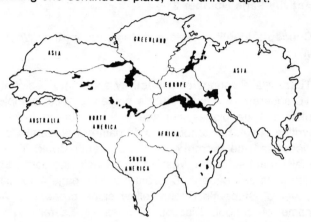

Answer(s): Youngsters should see the approximate "fit" between Figures 1 and 2.

PLANET EARTH SOMETIMES QUAKES...

Activity Page Number 11

Concept: Earthquake

Tip(s): Give one rock, with corner taped, to each youngster. Face same direction as youngsters. Turn rocks. Discuss quarter and half turns plus counterclockwise and clockwise. Do several practice activities before doing activity sheet. For an excellent pamphlet *Safety and Survival in an Earthquake* and related materials write to: U.S. Geological Survey, 604 South Pickett Street, Alexandria, VA 22304.

Answer(s): A = 3, B = 1, C = 4, D = 4

AND RUMBLES

Activity Page Number 12

Concept: Volcano

Tip(s): Introduce activity by playing "rumbling" music. Compare to sounds emitted by volcano such as Mt. St. Helens. Complete written activity. Discuss point of "contact" where magma contacts another surface and causes a change to take place, eg., sandstone changing to quartzite, limestone to marble, shale to slate or granite to gneiss. Building a volcano can be done as an individual or whole class activity. Be sure juice or film can is located inside a container that will catch lava. Bits of styrofoam can be made from discarded packing material, fast food containers or styrofoam trays.

Answer(s): (1) gas and dust (ejecta) (2) crater, (3) conduit, (4) side vent eruption, (5) lava flow, (6) liquid magma, (7) hot rock magma

GEOLOGICAL WONDERS OF THE WORLD

Activity Page Number 13

Concept: Rocks and minerals make up geological features

Tip(s): Introduce this long-term activity by having youngsters identify the world's seven major continents and three oceans. Teach youngsters how to read a map including its legend. Review geological map and compare to world map. Assign one geological wonder per student. Have youngsters find out as much information as they can about their feature and report their findings to the class.

Answer(s): Continents: A = Asia, B = South America, C = North America, D = Africa, E = Europe, F = Antarctica, G = Australia. Oceans: H = Pacific, I = Atlantic, J = Indian. Geological features: (1) Rosetta Stone, (2) Stonehenge, (3) Great Pyramid of Cheops, (4) White Cliffs of Dover, (5) Blarney Stone, (6) Rock of Gibraltar, (7) Mt. Everest, (8) Dead Sea, (9) Lake Eyre, (10) Mariana Trench, (11) Victoria Falls, (12) Tugela Falls, (13) Angel Falls, (14) Great Wall of China, (15) Gobi Desert, (16) Sahara Desert, (17) Plain of Jars (Laos), (18) Angkor Wat, (19) Atlantis (possible location), (20) Ayers Rock, (21) Cuzco, (22) Tikal, (23) Sangay Volcano, (24) Mt. Fuji, (25) Grand Canyon, (26) Cape Horn, (27) Weddell Sea

GEOLOGICAL WONDERS OF THE UNITED STATES

Activity Page Number 14

Concept: Rocks and minerals make up geological features

Tip(s): Similar to Activity Number 13 above, this activity can be done as an individual, small or large group activity.

Answer(s): Completed activity sheet looks like this:

(1) San Andreas Fault, (2) Mt. St. Helens, (3) La Brea Tar Pools, (4) Badlands, (5) Petrified Forest, (6) Grand Canyon, (7) Old Faithful, (8) Yosemite Falls, (9) Rocky Mountains, (10) Carlsbad Caverns, (11) Niagara Falls, (12) Great Salt Lake, (13) Lake Okeechobee, (14) Cape Cod, (15) Meteor Crater, (16) Cleopatra's Needle, (17) Great Smoky Mountains, (18) Everglades, (19) Mackinac Island, (20) Hot Springs, (21) Yellowstone, (22) Jackson Hole, (23) Mt. McKinley, (24) Hawaii Volcano House, (25) Crater Lake, (26) Death Valley, (27) Glacier National Park, (28) Mammoth Cave, (29) Wisconsin Dells, (30) Stone Mountain

FACING ROCKS NEAR YOUR COMMUNITY...

Activity Page Number 15

Concept: Characteristics of rock formations

Tips(s): Introduce activity by having youngsters try to identify the four faces found on Mt. Rushmore. Develop the concept that rocks too have faces. Have youngsters complete the activity. Be aware of safety precautions when near road cuts and areas of construction.

Answer(s): Part I—Answers will vary. Part II—(1) George Washington, (2) Thomas Jefferson, (3) Theodore Roosevelt, (4) Abraham Lincoln, Mt. Rushmore, South Dakota

AND DOWNTOWN

Activity Page Number 16

Concept: Uses of rocks and minerals in the construction of buildings

Tip(s): This activity should be done with adult permission and supervision from a parent or family member. If parents are used as leaders, a note that emphasizes safety can be sent home along with a copy of the activity. If a trip downtown cannot be arranged, the teacher can take photographs or slides of buildings for classroom activity. Distribute samples of rocks found on downtown buildings to youngsters. Discuss the importance of each. Funeral directors often can obtain free samples of rocks used in tombstones from dealers. See Activity Number 25 for correlated activities.

Answer(s): Answers will vary.

ROCKS AND MINERALS AT WORK...

Activity Page Number 17

Concept: Metals, found in rocks, are used to make things.

Tip(s): Introduce activity by having youngsters conduct a scavenger hunt to identify things found in the classroom that are made from metals. Completed activity shows where these metals came from and how they were made.

Answer(s): Answers will vary.

AND AT SCHOOL

Activity Page Number 18

Concept: Rock cycle

Tip(s): Introduce activity with a discussion of an exercise cycle or bicycle used to maintain fitness. Note how heat and pressure (brakes) are used in these cycles. Discuss how rocks, too, go through cycles of heat and pressure. With older youngsters, discuss hydration, carbonation and oxidation of rocks. Complete activity. Discuss other life cycles such as plant and animal life.

Answer(s): 1. Magma, 2. Igneous rocks, 3. Sediments, 4. Sedimentary rocks, 5. Metamorphic rocks to Magma. (Stop 1), 6. BONUS: Sediments by Weathering

TINY ROCK CRYSTALS MAKE . . .

Activity Page Number 19

Concept: Crystals take a definite geometric shape

Tip(s): Introduce activity by sprinkling different crystals on an overhead projector. Note size, shape, and color. Note how crystals are also found in various rocks such as a geode. Have youngsters complete activity. Note how crystals come in many sizes and shapes, including the shapes of a diamond and square. Excellent reference book is *Puzzles in Space: Crystals to Make and Explore* by David Stonerod, Stokes Publishing Company, P.O. Box 415, Palo Alto, CA 94302

Answer(s): Salt crystals have square, flat sides that, when put together, take on a cubical shape. Diamond crystals have four equal sides, two equal obtuse angles (greater than 90° but less than 180°) and two equal acute angles (less than 90°). A pattern to use for a diamond-shaped crystal looks like this:

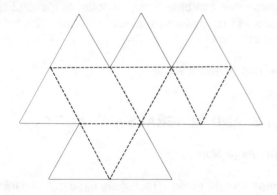

LARGE CRYSTAL GARDENS

Activity Page Number 20

Concept: Crystal garden

Tip(s): Make up a stock solution of the mixture. Add to coal each day as needed to help crystals continue to grow. Place newspaper or plastic under container that holds the crystal garden. Alum is also an excellent medium to use for growing crystals. Caution youngsters not to eat alum.

Answer(s): Growth of a crystal garden.

CLASSIFY ROCKS INTO . . .

Activity Page Number 21

Concept: Three major classes of rocks: Igneous, Metamorphic and Sedimentary

Tip(s): Do this activity in conjunction with Activity Number 22. Laminate cards for lasting durability.

Answer(s): Igneous—Granite (F), Basalt (I), Obsidian (R), Diorite (E), Secret Word = Fire. Metamorphic—Slate (H), Schist (E), Marble (A), Quartzite (T), Secret Word = Heat. Sedimentary—Conglomerate (W), Sandstone (O), Limestone (R), Shale (N), Secret Word = Worn.

A POCKETFUL OF ROCKS

Activity Page Number 22

Concept: Three major classes of rocks: Igneous, Metamorphic and Sedimentary

Tip(s): Pockets can be mounted easily on poster board or tagboard, or large pockets can be drawn on poster board. Mount paper clips on back of pictures of rocks. Mount small strips of magnetic tape to board. Youngsters mount correct picture of rock on matching pocket.

Answer(s): Cards should be placed in pockets according to answers found in Activity Number 21.

COLLECT AND TEST ROCKS

Activity Page Number 23

Concept: Classification of rocks based on rock characteristics

Tip(s): Introduce activity by having youngsters sit in a circle. Have each youngster place one of his shoes in the center of the circle. As a group, classify the shoes in as many ways as possible. Youngsters use same method to classify their five collected rocks.

Answer(s): Answers will vary. Cards should follow the order of rocks found on the Mohs' Scale of Relative Hardness.